SEE HOW PLANTS GROW
Fruits

Nicola Edwards

PowerKiDS
press.

New York

Published in 2008 by The Rosen Publishing Group, Inc.
29 East 21st Street, New York, NY 10010

First Edition

The publishers would like to thank the following for allowing us to reproduce their pictures in this book:
Alamy: 14 (Darren Matthews). Corbis images: cover and 4 (Jutta Klee). Ecoscene: 7 (Norman Rout). Eye Ubiquitous: 11 (Gavin Wickham). Garden Picture Library: 22 (Botanica). Getty images: 6 (Peter Lilja), 8 (Michelle Quance), 13 (Michael Orton), 15 (Howard Rice), 16 (Dave King), 17 (Johner), 18 (Martin Harvey), 19 (Hein von Horsten), 20 (photos alyson), 21 (Richard Nowitz), 23 (Key Sanders). Wayland Picture Library: title page and 10, 5, 9, 12.

Library of Congress Cataloging-in-Publication Data

Edwards, Nicola.
  Fruits / Nicola Edwards. -- 1st ed.
    p. cm. -- (See how plants grow)
  Includes index.
  ISBN-13: 978-1-4042-3701-8 (library binding)
  ISBN-10: 1-4042-3701-1 (library binding)
  1. Fruit--Juvenile literature. I. Title.

QK660.E39 2007

581.4'64--dc22

Manufactured in China

Contents

What Are fruits?

Fruits grow on flowering plants. They contain **seeds** that will grow into new plants. Fruits come in many different shapes, sizes, and colors.

They may have smooth, shiny skins, thick peels or rough, hairy surfaces. They vary a lot in smell and taste.

▼ Many trees are flowering plants and produce fruits such as apples.

Fruits are important food for people around the world. They contain **vitamins**, which we need to stay healthy.

▼ How many types of fruits can you see?

Different kinds of fruits

You might think of a fruit as something you'd eat in a fruit salad, such as a grape, melon, or mango. A fruit is something that contains seeds, so there are many things in nature that can be described as fruits.

▲ Coconuts are the fruits of the coconut palm tree.

Fruit Fact

Not all fruits can be eaten. In fact some are **poisonous**, including many berries.

Fruits include things you might not have thought of, such as pea pods, tomatoes, and eggplants.

See the fruits growing in these eggplant pots.

Where do fruits grow?

If you walk around parks and gardens, you may see many different fruits. Summer and fall is when many fruits grow.

▶ These strawberries grow on a farm where visitors pick fruits fresh from the plant.

Some people grow fruits, such as apples and pears, on trees in their yard, or plant beans, zuccchini and peas. Others have greenhouses in which they grow fruits, such as tomatoes and peppers that need warmer **conditions**.

▲ Workers on this farm in France grow apples to sell in many different countries.

Fruits around the world

Fruits grow all over the world, except in the areas near the North and South Poles where it is too cold. Many fruits, such as papayas and bananas, grow well in **tropical rain forests**, where it is very warm and wet. Other fruits, such as lychees, need plenty of rain.

▼ Oranges, lemons, and limes are **citrus fruits**. Citrus fruits grow well in warm sunshine.

▶ These bananas are growing on a tree in a Caribbean rain forest.

Starting from a seed

Some fruits contain many seeds, others have just a few seeds, and some surround a single large seed. Each seed is the start of a new plant. A seed has a hard outer casing. When the seed begins to grow, the casing splits open.

▼ Look inside these fruits. Some contain more seeds than others.

▼ A shoot is beginning to grow from this pea seed.

shoot

seed

The parts of a plant

The **roots** of a fruit plant spread through the soil, anchoring the plant in the ground. The roots take in water and **nutrients** from the soil. These travel via the **stem** to all the parts of the plant. The stem supports the leaves and flowers that develop from **buds**.

▼ You can see the developing roots and stem of this **snow pea** plant.

Fruit Fact

Some fruits, such as grapes, grow on climbing plants called **vines**.

Leaves make food for the growing plant. Flowers use their bright petals and strong scents to attract insects and other animals.

leaves

buds

flowers

15

From flower to fruit

The flowers of a plant produce **pollen**. When insects land on the flowers, they **fertilize** the plant by moving pollen from one part of the flower to another. The petals then fall and a fruit begins to form.

fruits

flower

Fruit Fact

Some fruits, such as pears and peaches, become softer and sweeter as they ripen.

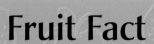 You can see the strawberries beginning to grow on this strawberry plant.

▼ These blackberries change from green to red to black as they ripen.

How fruits spread seeds

Fruits help plants to spread their seeds so that the seeds will grow into new plants. Fruit plants spread their seeds in different ways—by air, by water, and by animals, such as birds and chimpanzees.

▼ This coconut floated on the sea until it reached land and began to sprout.

Animals eat fruit because they are sweet and juicy. The seeds pass unharmed through the animals' bodies and fall to the ground in their droppings.

▶ Seeds in this bird's droppings could sprout into new plants.

How do we use fruits?

We eat fruits such as apples, plums, and apricots raw, or cook them in pies. Some fruits, such as bananas and cranberries, taste good when they are dried.

Fruit Fact

You can use lemon juice as invisible ink!

Eating fruit is a good way to keep healthy.

Dried fruits such as **gourds** can be made into musical instruments. Fruits, such as mangoes and papayas, are used to make scented bath oils and shampoos.

▼ Pumpkins can be carved into shapes to make Halloween lanterns.

Grow your own fruit

Grow tomatoes from a pack of seeds. Fill a seed tray with soil. Plant a seed in each section and cover it with a layer of soil. Water the soil to keep it moist. Tie a plastic bag over the tray, and put it somewhere warm and light. When the seedlings appear, move them to a separate pot or put them in a yard or window box.

▼ Follow the stages in your tomato plant's life.

Fruit Fact

When the tomato seeds start to grow, this is called **germination**.

How will you know when the fruits are ripe and ready to eat?

Glossary

buds
The parts of a plant in which leaves and flowers form.

carbon dioxide
A gas in the air that plants use to make food.

citrus fruits
Fruits such as oranges.

fertilize
When the male and female parts of a plant merge.

germination
How a seed starts to develop into a plant.

gourd
A fruit similar to a pumpkin or a squash.

nutrients
Food taken in by the roots of a plant from the soil.

pollen
A yellow dust that a female part of a flower needs to form a seed.

roots
The parts of a plant that hold it in the soil and suck up water and nutrients.

seeds
The parts of a plant from which new plants develop.

shoot
The start of upward growth from a seed.

snow peas
A type of pea, but the pod is eaten too.

stem
The part of a plant that holds it upright.

tropical rain forests
Forests that grow in areas where it is hot and wet.

vitamins
Substances found in fruits that help us keep healthy.

Web Sites

Due to the changing nature of Internet links, PowerKids Press has developed an online list of Web sites related to the subject of this book. This site is regularly updated. Please use this link to access this list: www.powerkidslinks.com/shpg/fruit/

Index